Drunk Science

What happens to our bodies as we drink

By: Dr. Mike Cameron

Dedicated to my beautiful daughter, Aria. You're the reason why daddy drinks.

Disclaimer

(AKA I am not getting sued if one of you goobers gets alcohol poisoning from trying to recreate the symptoms in this book)

I get it. A lot of people think drinking booze is cool. That fact, combined with targeted marketing, and peer pressure (usually from assholes, who aren't your friends to begin with if you think about it), make it even more irresistible. The point of this book is not to tell you to stay away or gravitate towards drinking; it is to educate you on what happens to your body when you drink so that you can make an informed decision. A lot of times, I will try and use humor (poorly) to keep your attention, but just know that alcohol, like any other controlled substance, creates changes in the body that can become permanent if abused so it is something that commands respect. If you don't believe me, flip to the second to last chapter of the book where I talk about the variety of health problems that alcoholics can face on a daily basis.

Alright, on with the show!

Alcohol 101

Throughout this book, I will use terms like booze, hooch, firewater, etc. to refer to alcohol. In fact, the chemical name for this stuff is ethanol or ethyl alcohol. It is an organic compound, made up of carbon, oxygen, and hydrogen.

$$\begin{array}{c} \text{H} \quad \text{H} \\ | \quad \ \ | \\ \text{H}-\text{C}-\text{C}-\text{O}-\text{H} \\ | \quad \ \ | \\ \text{H} \quad \text{H} \end{array}$$

Alcohol- the cause of and solution to, most of life's problems

-Homer Simpson

Ethanol is created through a process called fermentation. This is where yeast is added to sugar water, the yeast eats the sugar for energy, and then gives off ethanol and carbon dioxide as by-products. It can be further purified by a process called distillation; in which the ethanol is separated from the water along with other impurities that hitch a ride during fermentation. Even though it is possible to make 100% pure ethanol, ALL alcoholic drinks do not come anywhere near that level due to the fact that drinking pure ethanol is a very efficient way of causing severe alcohol poisoning, and most likely, death. For this reason, most alcoholic drinks are watered down. For example, beer is usually 4-6% ethanol, wine is 7-15%, and spirits are 40-90%. Those numbers are important to know, because the higher the percentage, the quicker your body will freak out if you do not drink responsibly. This is probably why you don't see people beer bonging straight vodka.

Alcohol readily dissolves in water, which makes it very easy to be absorbed in our digestive system. On average, 20% is absorbed through our stomach, and the remaining 80% is absorbed though the duodenum (the first part of our small intestine). From our digestive system, alcohol is dumped directly into our blood. The level of ethanol in our blood at any given time is referred to as our blood alcohol concentration (BAC). Keep that in mind, because we are going to talk about that crap A LOT in this chapter.

Once the ethanol makes its way through our circulatory system (arteries and veins), it eventually makes its way to the liver. Our liver is the real MVP when it comes to keeping us from dying by alcohol poisoning. **This is the time to focus, because here comes some important shit.** Our liver contains an enzyme called alcohol dehydrogenase that breaks down ethanol to a compound called acetaldehyde. From there, another enzyme called acetaldehyde dehydrogenase breaks down acetaldehyde to acetic acid. Acetic acid is just a fancy term for vinegar, and easily can be broken down into water and carbon dioxide by the body. The water then gets peed out, and the carbon dioxide is expelled though normal breathing.

For all you visual learners:

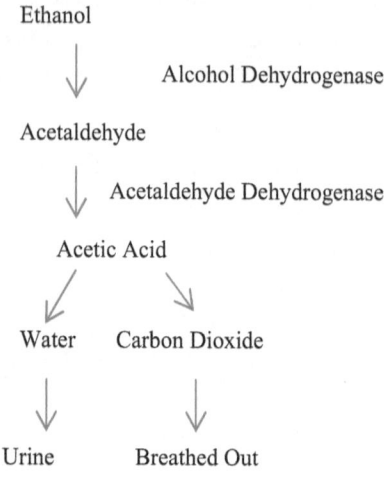

The speed of the breakdown of ethanol is different from person to person, and is due to various genetic and physiological factors. Some people have more enzymes than others that speed up the breakdown of ethanol. Long term alcoholism also causes increased levels of both alcohol dehydrogenase and acetaldehyde dehydrogenase, which makes it harder to get drunk. Healthy livers also breakdown ethanol faster than those of older, more damaged livers. The longer the liver takes to break down alcohol, the more that it backs up into the circulatory system. When this happens, it causes that blood alcohol content (BAC) level to rise. When your BAC reaches certain levels, we experience various symptoms:

BAC = 0.03-0.12% (by the way, impaired driving is BAC that is 0.08% or above)

This is the happy phase of drinking. You experience a slight sensation of euphoria and self- confidence rises. Unfortunately, your judgement and decision making skills start becoming impaired as well.

BAC = 0.09-0.25%

EVERYTHING IS EXCITING FOR NO PARTICULAR REASON! However, you won't really remember being excited because in this phase you start having trouble with your memory, along with other problems such as loss of balance, blurred vision, and decreased reaction time.

BAC = 0.18-0.30%

We are now entering the not so fun stages. Confusion usually sets in. Speech starts slurring, and walking now becomes a chore.

BAC = 0.25-0.4%

At this level, it is common to pass out. Not that you could, but standing and walking are just about impossible now. If that wasn't enough, your body decides that all that booze in your system has got to get the hell out, so you start to vomit.

BAC = 0.35-0.5%

At this point, you should be in the hospital. It is very possible to become unresponsive and slip into a coma. Heart rate and breathing rate both significantly drop. Medical intervention is a necessity at this point (get ready to have that stomach pumped).

BAC = more than 0.5%

Organ systems start shutting down, so in short….you dead.

So there is a quick breakdown of what happens when you drink, but how the hell does all that happen? Well I'm glad you asked. Our body is filled with little guys called neurotransmitters. These are chemicals that nerves use to cause certain actions to happen. Ethanol screws with two important neurotransmitters. The first one is called gamma-Aminobutyric acid, or GABA for short (because I'm not getting paid by the word to write this book). GABA is an inhibitory neurotransmitter, meaning whenever it is released, it decreases the function of anything it touches. The other neurotransmitter is called Glutamine. This one does the opposite of GABA by making everything it touches more hyper than a puppy on cocaine. All the changes mentioned above happen because different organs, glands, and tissues become affected by either GABA or Glutamine (or sometimes both).

So there you go. The basics of ethanol, and how it works. If you just want to stop here, you will still sound pretty damn smart when you talk about this in the future. However, if you really want to blow people's minds, keep reading because the rest of the book deals with how each symptom of drinking actually happens.

Breaking the Seal

So you decided to have a few drinks, and the next thing you know you now have to pee. The only thing is, after you pee, you find yourself going back to pee about every 20-30 minutes. This fun little phenomenon is sometimes referred to as "breaking the seal". I know it's a clever little name, but in all actuality, it is down playing the fact that ethanol has reached the inner part of your brain and is starting to mess with shit.

In the middle of our brains we have a structure called the hypothalamus. It is like the boss of most of our hormones and helps keep everything in check. Remember when I said ethanol increases the production of GABA (if you don't, go back to the previous chapter, read it, and start paying better damn attention)? Well GABA attaches to receptors on the hypothalamus and slows it way the hell down. This is important because one of the many jobs of the hypothalamus is to talk to another important structure called the pituitary gland (found directly underneath the hypothalamus in the brain). Not only is GABA slowing down the transmission of information from the hypothalamus to the pituitary, but that asshole is also slowing down the pituitary gland itself. Now here's the important part – one of the many hormones made by the pituitary gland is called anti-diuretic hormone. The whole purpose of this hormone is to make us retain water (see where I'm starting to go with this?). This hormone works by talking to the cells of the kidneys (known as nephrons) and telling them to not filter out water from the blood. However, since the hypothalamus as well as the pituitary gland are slowed by GABA, the levels of anti-diuretic hormone reach incredibly low levels, meaning that you can no longer retain water like you should, so you start to pee a lot. This cycle usually does not stop until GABA levels return to normal (meaning you stop drinking).

This whole thing may seem tame, but in case you didn't know, humans can't survive without water. It's kind of a big deal. Keep this in mind when we go into how hangovers work.

Boozes Makes Brain No Think Good

Well the constant trips to the bathroom to pee haven't stopped you from drinking, so let's talk about the next stage where it starts affecting your brain. Due to your rising BAC level, your concentration of GABA in your blood starts to rise as well, and starts to stick to every damn thing. This is what accounts for a lot of the common symptoms associated with drinking. For example, slurring your speech and having double vision are two common problems. The speech problem occurs due to the fact that GABA slows down the transmission of impulses from various nerves in our brain to the muscles of our lips and tongue. Once this happens, we can't form words as quickly as our brain wants to, so they all kind of blend together. The same thing happens with our vision. We have muscles in and around our eyes that help us to focus. When GABA concentrations get too high, the nerve impulses slow down which, in turn, slows down the speed at which our eyes focus.

If those were the only two things that occurred, becoming drunk wouldn't be too horrible, unfortunately, booze messes a whole bunch of shit up in your brain at the same time. GABA slows down the function of your cerebellum, which is the back portion of your brain responsible for balance and coordination. So congratulations, you now have trouble walking and performing complex movements (which is why police make you do that stand on one foot and touch your nose thing).

At this point, you can't really talk or see well, and it's even hard to walk straight, but the good news is that you most likely won't remember it! GABA inhibits the function of a structure in our brain called the hippocampus. The job of the hippocampus is to form short term memories, as well as assist in learning. So that is why it's very common to wake up the next day with no recollection that you decided to drunk dial your ex about fifty times. Since we are on this topic, what neurotransmitter do you think is responsible for lowering our inhibitions, which gives us all these "great" ideas? That's right! It's GABA! That son of a bitch messes everything up! GABA also acts on another part of our brain called the amygdala. This little guy is responsible for emotional memory as well as survival instincts. So you can blame your amygdala when your ex calls you a creep and blocks your number.

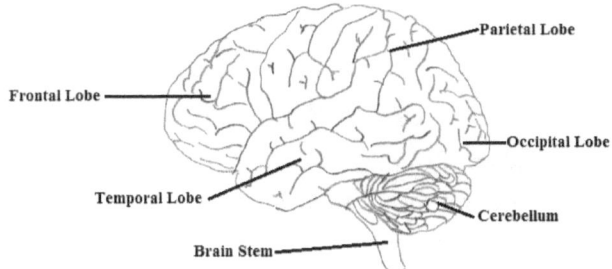

Outside of the brain (I'm a scientist, not an artist)

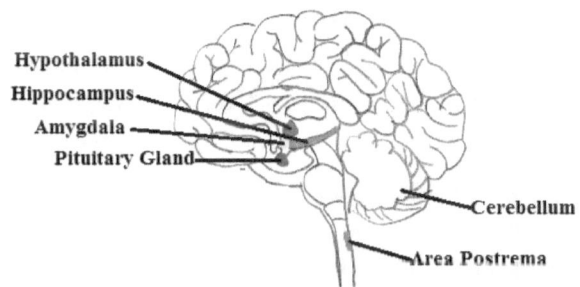

Inside of the brain (did I mention I'm not an artist)

Everybody outta the pool!

To recap, at this point in your drinking adventure, you can no longer talk coherently, it's hard to see, stand, walk, and remember things….and, oh yeah, you constantly have to pee. If, after all this, you still decide to keep drinking, you get to experience the exciting final stages of excessive drinking; vomiting and blacking out.

If you reach this point, your BAC gets dangerously close to toxic levels. Luckily, our brain has a structure called the area postrema which acts like a sensor, scanning our blood for toxins. When the ethanol levels in our blood get too high, the area postrema activates, causing us to vomit. Funny enough, that's not the only way excessive drinking can cause you to toss your cookies. Ethanol is an irritant, and having too much in your stomach and intestines can cause spasms, and vomiting. The third way involves going through the "spins". At the innermost part of our ear is a structure called the semicircular canals. Inside of this guy is a gel like substance known as endolymph. Embedded in the endoplymph, and attached to the walls of the canals are little hairs called stereocilia. Normally, when we move our head, it causes the endolymph to slide, and pull on the hairs, which our brain uses to interpret what direction the head is pointed. Ethanol thins our blood, which causes a density difference of the endolymph in the semicircular canals. This change in density causes the little hairs to get confused, and tell the brain that the head is moving, when it is not. This will make you vomit due to the sensation that you are spinning in circles, hence, the "spins". SCIENCE!

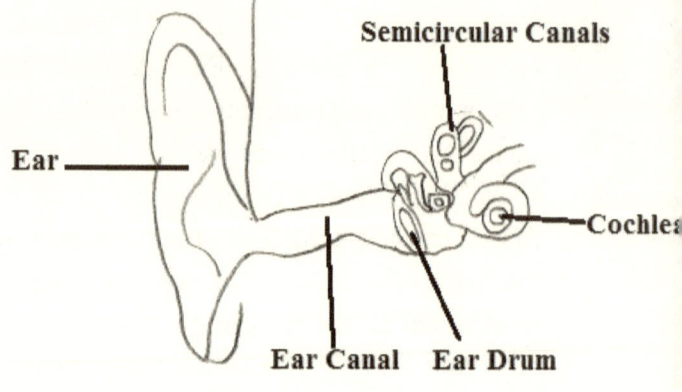

That's supposed to be your ear (don't judge)

As for the blacking out part, it's just a continuation of what we already talked about when you lose short term memory (GABA is reducing the activity of the hippocampus). However, now that the levels of ethanol in your blood have become incredibly high, the hippocampus is almost completely shut down, and can't form new memories. At this stage you can either experience fragmentary memory loss (forgetting small periods of time), or en bloc memory loss (forgetting large periods of time). It just depends on how severe the BAC levels get. Neither of which are really a good thing.

These final stages are extreme because they are protecting you from dying. If your GABA levels get too high, they will start affecting your heart and lungs. GABA slows things down, remember? So too much of it will slow down your heart and breathing rates to dangerous levels, which can most certainly kill you. No humor in this part. It's serious. Don't get to that point.

The Aftermath

If getting drunk was how people forgot they were mortal, then hangovers were how they remembered.

-Matt Haig

Well now you did it. You won the contest of over drinking last night, and your award is that you get to wake up with a hangover. This entails a wide variety of symptoms including headaches, nausea, diarrhea, muscle soreness, fatigue, and just a general sense that you hate yourself from last night.

Remember in the initial stages of drinking when I talked about having to pee a lot? Well here is where that becomes something important. See that never stopped. You keep losing water when you drink because of the decreased ADH (anti diuretic hormone) levels in your body. Water plays a vital role in many bodily functions and with severe dehydration comes severe symptoms.

Blood is mostly water, so when our water levels drop, our blood volume drops as well. When this happens, our body tries to maintain a normal blood pressure by causing our blood vessels to constrict. Narrowing the blood vessels can maintain our blood pressure, but since the diameter of the blood vessels are reduced, it will not let as much blood through (think walking in a single file line instead of a crowd of people walking together). This results in tissue not getting essential nutrients as fast, including oxygen. When this happens in the brain, the blood vessels dilate as a way to try and deliver more oxygen to itself. This process can cause minor swelling, which results in a lovely pounding headache.

Besides causing dehydration, the breakdown of ethanol itself causes the formation of acetaldehyde (remember from one of the first chapters). If we drink to excess, our livers can't keep up with the breakdown of it, causing the acetaldehyde to spill into our blood stream. This chemical is extremely irritating to just about everything it touches. It inflames mucous membranes that line our digestive system which can cause acid reflux, nausea, stomach aches, and diarrhea. Another fun fact about acetaldehyde is that it is highly toxic. So during a hangover, you literally have poison in your blood, and since you are dehydrated, the poison becomes concentrated. Starting to understand why you feel like shit the next day?

Over the years, students and patients have asked me if there are ways to cure a hangover. Now there are things that you can do to help your body get back to normal (like drink water), but overall, time is the only thing that will cure it. Our bodies need time to expel the toxin from out blood, as well as all the tissues that the acetaldehyde affected. Those with a faster metabolism may experience quicker recovery times from a hangover, but there is no magic bullet that instantly cures a hangover. The best way to cure a hangover is to never get one. I know that sounds like bullshit fortune cookie advice, but it's true. Like I said in the beginning of this book, my job is to educate you on what happens when you drink. It is up to you to recognize when you are experiencing the symptoms of drinking too much, and try to make the best decision possible for yourself.

To drink, or not to drink. That is the question.

At this stage in our little journey, you may be calling me a hypocrite. I have said several times that the point of this book is to educate you, not direct you, in your decision making process. Yet everything so far has been negative. The reasoning behind that is plain old biochemistry. Ethanol and some of its metabolites are toxic, plain and simple. Like most things that are toxic, if you consume a lot of it, bad shit happens. Alcohol is no exception. If something contains ethanol, than your body will not like it if you consume a lot of it.

However, there is evidence to support the fact that certain ethanol containing drinks do actually have health benefits if drank in **MODERATION**. Moderation is the key word in that last sentence. That's why I capitalized and bolded that shit.

Wine made from dark red grapes contains a chemical called ellagic acid. This acid has been shown to help manage obesity by increasing liver metabolism.

Antioxidants in wine can also increase levels of nitric acid in the blood which relaxes the muscles in the walls of the arteries so that the arteries can widen, causing more blood and nutrients to pass through. There is also a chemical in wine called resveratrol, which is thought to protect against things like cancer, heart disease, and memory loss. Now did I just say that drinking wine will make you skinny, and cancer free with an excellent memory? No, don't be stupid. I'm saying that research exists that states moderate consumption of wine actually has some health benefits (by the way, in most of the research, moderation means a glass a day).

Let's not leave beer out. There is plenty of research that proves moderate consumption of beer can be fairly healthy as well. Beer contains a chemical called xanthohumol that may protect brain cells from the damaging effects of certain diseases such as Parkinson's and Alzheimer's. It can also decrease the risk of getting kidney stones (so can just drinking water, but this book isn't about water). Also, beer has been shown to maintain healthy levels of homocysteine in our body, which can decrease the chance for atherosclerosis and blood clots. Besides the various chemicals in it, beer also contains vitamins and minerals. Most beers are high in B vitamins, magnesium, calcium, and silicon, which can aid in a lot of metabolic functions as well as increasing bone density. Dark beers are high in iron, which is a key component in making hemoglobin (the major protein that helps hold onto oxygen in a red blood cell).

Liquors do not contain a lot of health benefits due to the fact that they undergo distillation, and almost everything but the ethanol is removed. So I guess those hillbillies on TV that make moonshine must be staying healthy some other way.

The research studies that put a positive spin on drinking may differ from one another in many ways, but the one thing that they have in common is that the subjects in every single study only drank a moderate amount of alcohol. Drinking to excess not only negates the health benefits listed above, but it allows toxic ethanol to build up in our bodies.

If you decide to drink way too much for way too long, your body will start getting pissed off. Our organs are basically like parts of a machine, and if we over work them, the parts break. Our livers will become scarred, and develop a condition called cirrhosis, neurons in our brain start dying off, causing memory loss, as well as making it much harder to learn new things. The mucous in our digestive system decreases, which can lead to irritation, and conditions ranging from heart burn to ulcers to not absorbing vitamins and nutrients properly. Our heart rate increases, which makes the heart work faster, causing premature aging of it. Even the function of the pituitary gland can decrease. Besides making anti-diuretic hormone, it also makes other things like follicle stimulating and luteinizing hormones. These guys play an important role in the creation of testosterone/sperm in men and estrogen/eggs in women, so you can even get fertility issues with long term alcohol abuse. Oh yeah, let's not forget that alcohol abuse has been tied to higher rates of getting cancer. So that's a biggie.

As cynical as I can come off, I do believe that most people want to be smart, and do not actively try to kill themselves. This is what makes ethanol such an asshole. When GABA interacts with a part of our brain called the limbic system, it can trigger depression, but at the same time it causes the brain to increase production of dopamine which activates our reward centers. This tricks us into thinking we are feeling good, when in fact we are depressed. The effects of dopamine are short lived, and even decrease over time, so when it wears off, the depression kicks you in the face with a pair of steel toed boots. We then start to drink in order to get that dopamine boost, but in reality it just makes the depression even greater, and now you are stuck on a vicious cycle known as alcoholism. This is such an easy path to go down, and it usually starts by drinking socially for fun when we are young. I have seen people with alcoholism. I have friends who are alcoholics. It ruins lives.

So now what?

By now you have hopefully read through the entire book instead of just flipping to the second to last chapter like said in the disclaimer (that's a dick move, by the way). Your reward for enduring my bad jokes throughout this book is that you now should have better than average knowledge of how our bodies act when we drink, and what each symptom actually means. The hard part is deciding what to do with this knowledge. You most likely will come to a point where you will make a conscious decision to drink more, or to step away. When that time comes, use what you have learned in this book to help make an informed decision. Being educated is never a bad thing, and is what separates us from the monkeys (well, not those smart ones that know sign language, they have their shit together). The whole purpose of this book was to try and make your life just an ounce easier, and hopefully, with your help, it will work.

References

Ferk F, Huber W, Filipic M, Bichler J, er al. (2010). Xanthohumol, a prenylated flavonoid contained in beer, prevents the induction of preneoplastic lesions and DNA damge in liver and colon indiced by the heterocyclic aromatic amine amino-3-methyl-imidazo[4,5-f]quinolone(IQ). Mutation Research/Fundamentals and molecular Mechanisms of Mutagenesis. 2010; 691(1-2).

Galanter M. (2006). Recent Developments in Alcoholism. Vol 14. Springer Science and Business Media. Berlin, Germany.

Mayer O Jr, Simon J, Rosolova H, (2001).A population study of the influence of beer consumption on folate and homocysteine concentrations. European Journal of Clinical Nutrition. 2001; 55(7) 605-609.

Patel V. (2015). Molecular Aspects of Alcohol and Nutrition. Academic Press. Cambridge, MA.

Preedy V, Peters T. (2002). Skeletal Muscle, Pathology Diagnosis and Management of Disease. Greenwich Medical Media. Greenwich, CT.

Ramesh V, Hajime O, Pawan S, Nilanjana M (2006). Significance of wine and resveratrol in cardiovascular disease: French paradox revisited. Exp Clini Cardiol. 2006; 11(3):217-225.

Salamon A, Baca E, Baranowski K, Michalowska D. (2012). The evaluation of anti-nutritive components in beer on the example of oxalic acid. Rocz Panstw Zakl Hig. 2012; 63(1)37-42.

Shay, N. (2015). Another reason to drink wine: it could help you burn fat. Internet. https://today.oregonstate.edu/archives/2015/feb/another-reason-drink-wine-it-could-help-you-burn-fat

Sommez U, Sommez A, Erbil G, Tekmen I, Baykara B (2007). Neuroprotective effects of resveratrol against traumatic brain injury in immature rats. Neurosci Lett. 2007; 420(2):133-137.

Torres-Pérez M, Tellez-Ballesteros RI, Ortiz-López L et al., (2015). Resveratrol Enhances Neuroplastic Changes, Including Hippocampal Neurogenesis, and Memory in Balb/C Mice at Six Months of Age. PLOS One 22;10(12):e0145687. doi: 10.1371/journal.pone.0145687. eCollection 2015.

Tucker K, Jugdaohsingh R, Powell J, Qiao N, et al. (2009). Effects of beer, wine, and liquor intakes on bone mineral density in older men and women. American Journal of Clinical Nutrition. 2009; 89(4)1188-1196.

Cover photo by Amanda Register

www.ingramcontent.com/pod-product-compliance
Lightning Source LLC
Chambersburg PA
CBHW031513210526
45464CB00007B/2897